CONTENTS

THE OBSERVER'S REALITY: UNLOCKING THE HIDDEN POWER OF YOUR ATTENTION

By

Tony Yustein © 2024
https://thecode.wiki

INTRODUCTION:
SEEING THE UNSEEN

Imagine this: a single act of observing something changes its behavior. At first glance, this might seem like a trick of perception—a psychological phenomenon. But in the strange, subatomic world of quantum mechanics, this is a fundamental reality. The mere act of observing can alter the state of a particle, transforming possibilities into certainty. This is the **observer effect**, a mind-bending concept that challenges our understanding of reality itself.

• •

The Observer Effect: What It Is and Why It's Mind-Blowing

The observer effect reveals an astonishing truth: at the quantum level, reality doesn't solidify until it's observed. In classical physics, we're accustomed to the idea that objects exist independently of whether we look at them. A rock is a rock, whether or not you see it. But in quantum mechanics, particles like electrons or photons exist as a haze of probabilities—like a cloud of potential positions or states—until an observer interacts with them. Observation "collapses" these possibilities into a single, definite outcome.

This concept is beautifully demonstrated by the famous **double-slit experiment**. When scientists shine a beam of particles, such as electrons, at a barrier with two slits, the particles behave like waves, creating an interference pattern on the screen behind the barrier. But when the particles are observed passing through the slits, something extraordinary happens: the interference pattern disappears, and the particles behave like solid objects. It's as if the act of observation forces them to "choose" a definite path. How could something so simple—just watching—change the nature of reality?

· ·

A Simple Introduction to the Quantum World:
Particles, Waves, and Probabilities

To grasp the observer effect, we must step into the quantum world —a place where the rules of everyday life don't apply. In this realm:

- **Particles act like waves**: Imagine throwing a pebble into a pond and watching ripples spread out. Quantum particles behave like those ripples, existing in multiple places at once until observed.

- **Probabilities rule**: Instead of knowing exactly where a particle is, quantum mechanics provides a "probability cloud" of where it might be.

- **Entanglement defies logic**: Particles separated by vast distances can be mysteriously connected, instantaneously sharing information.

If this sounds surreal, you're not alone. Even Einstein struggled with the implications of quantum mechanics, famously calling it "spooky action at a distance." Yet, quantum theory has proven itself through countless experiments and is the foundation of modern technology, from semiconductors to lasers.

· ·

The Central Question: How Does Observation Create Reality?

The observer effect raises profound questions: Does reality exist independently of observation? Is the observer merely a passive witness, or an active participant shaping what's observed?

This question isn't just for physicists—it's deeply personal. In your own life, how might your attention, focus, and awareness shape the outcomes you experience? Could the principles of quantum mechanics extend beyond particles to influence relationships, career paths, or even health? The observer effect isn't just about electrons; it's a lens through which we can explore our role in creating the reality we live in.

• •

A Roadmap for the Book

In the chapters ahead, we'll embark on a journey to explore the observer effect and its implications for science, philosophy, and daily life. Here's what to expect:

1. **The Quantum Basics**: We'll break down quantum mechanics into simple, everyday terms, including the famous experiments that revealed the observer effect.

2. **The Observer in Daily Life**: We'll explore how the observer effect mirrors phenomena in psychology, relationships, and self-improvement—showing you how to harness it in your own life.

3. **Science Meets Spirituality**: We'll dive into ancient wisdom and philosophical traditions that resonate with quantum principles, like the idea that consciousness plays a central role in shaping reality.

4. **Practical Applications**: You'll learn techniques to use the observer effect to transform your life, including mindfulness practices, attention-training exercises, and ways to "collapse" the potential into the possible.

5. **The Bigger Picture**: Finally, we'll consider the future of quantum observation, from cutting-edge technologies like quantum computing to the implications of living in a universe shaped by observation.

By the end of this book, you'll not only understand the observer effect but also feel empowered to use it as a tool for transformation. After all, if observation shapes reality, then you are not a mere passenger in life—you are its creator.

• •

CHAPTER 1: WHAT IS THE OBSERVER EFFECT?

The observer effect is one of the most fascinating and, frankly, bewildering ideas in modern science. Simply put, it's the concept that observing something can change its behavior. While this might sound like a quirky psychological phenomenon, in the world of quantum mechanics, it's a fundamental truth about how reality operates at the smallest scales.

To help us dive into this concept, let's start with a simple analogy:

• •

A Watched Pot Never Boils

Imagine standing over a pot of water on the stove. You're waiting, staring at it, willing it to boil. Minutes crawl by, and nothing happens. The saying goes, "A watched pot never boils." Of course, the water will boil eventually, but doesn't it feel like staring at it delays the process?

This feeling is the essence of the observer effect—your attention seems to change the experience of the pot boiling. While in everyday life, this is a psychological quirk, in quantum mechanics, observation has a much deeper and more literal impact.

• •

Observation Influences Outcomes

In the quantum world, particles like electrons and photons don't behave like tiny billiard balls, fixed and predictable. Instead, they exist in a state of **superposition**—a cloud of probabilities. These particles could be here, there, or even everywhere within a range, all at once. But when an observer measures or interacts with a particle, something extraordinary happens: the superposition collapses into a definite state. The act of observation determines where the particle is or how it behaves.

In other words, until someone looks, the particle doesn't commit to being in any one place. It's as though the universe is waiting for a witness before deciding what's real.

•••

The Famous Experiment: Schrödinger's Cat (Simplified)

To explain this idea, physicist Erwin Schrödinger famously created a thought experiment involving a cat. Imagine a cat in a sealed box with a mechanism that can randomly release poison based on the decay of a quantum particle. Until someone opens the box, the cat exists in a strange limbo—it's both alive and dead at the same time. Only when the box is opened, and the cat is observed, does it "choose" one fate or the other.

While Schrödinger didn't intend this to be taken literally, the analogy illustrates how observation in quantum mechanics forces a system to settle into a specific state.

•••

Quantum Physics and the Observer Effect

The observer effect is most famously demonstrated in the **double-slit experiment** (we'll explore this in detail in Chapter 3). In this experiment:

1. Scientists fire particles (like electrons) at a barrier with two slits.

2. If no one observes the particles, they behave like waves, creating an interference pattern on the screen behind the slits. This means the particles are in multiple places at once.

3. But if someone observes the particles as they pass through the slits, they behave like particles, not waves, and the interference pattern disappears.

This astonishing result shows that observation changes the behavior of quantum particles. It's as though reality itself is influenced by whether or not someone is watching.

. . . .

The Observer Effect Beyond Science

While the observer effect is a quantum phenomenon, its implications ripple far beyond physics. It invites us to consider profound questions:

- Does reality exist independently of observation, or do we play an active role in shaping it?

- If observation changes outcomes at the quantum level, could this principle extend to our daily lives? Could paying attention to something—like our goals or relationships—alter their trajectory?

We'll explore these questions in detail throughout this book. But for now, the observer effect introduces us to a mind-blowing possibility: **we are not passive witnesses to reality. We are participants in its creation.**

. . . .

Summary

The observer effect is the idea that observation influences outcomes. In quantum mechanics, this means particles exist in a cloud of probabilities until observed, at which point they take on a definite state. From the famous analogy of Schrödinger's cat to the eye-opening results of the double-slit experiment, the observer effect challenges our understanding of reality. Beyond the quantum world, it raises intriguing possibilities for how our attention and focus might shape the world around us.

In the next chapter, we'll dive deeper into the strange and wonderful rules of the quantum world to understand how particles and probabilities work. By breaking these ideas down step by step, you'll begin to see how the seemingly distant and abstract world of quantum mechanics connects to your life in surprising and meaningful ways.

CHAPTER 2: THE STRANGE WORLD OF QUANTUM MECHANICS

Quantum mechanics is a realm where the rules of the everyday world don't apply. Instead of the predictable, clockwork-like universe we're used to, the quantum world is governed by uncertainty, probabilities, and strange dualities. To make sense of the observer effect, we first need to understand these basics. Don't worry—while the concepts may sound intimidating at first, we'll break them down into digestible pieces, using relatable examples and simple analogies.

• •

Particles, Waves, and the Dual Nature of Reality

At the heart of quantum mechanics lies a surprising discovery: particles, the tiny building blocks of the universe, don't behave like tiny billiard balls as we might imagine. Instead, they exhibit a dual nature, acting both as particles and as waves.

1. **What's a Particle?**

   – Think of a particle as a tiny point, like a speck of dust. It has a specific location and size, and you can picture it as being "here" or "there."

2. **What's a Wave?**

   – Waves are all about movement and patterns, like ripples

spreading out on a pond. A wave doesn't have a single location; instead, it's a pattern of probabilities.

3. **The Quantum Twist: Particles as Waves**

– In the quantum world, particles like electrons and photons don't stay put. Instead, they spread out like waves, existing in a "cloud" of possibilities until something (or someone) interacts with them.

• •

Visualizing Wave-Particle Duality

Imagine you drop a pebble into a calm pond. The ripples that spread out are like the "wave" nature of a quantum particle. Now imagine if you could somehow freeze those ripples and say, "Here's where the pebble is." That's what happens when we observe a quantum particle—it "chooses" a specific location from the many possibilities represented by its wave.

• •

The Probability Cloud: Where Is the Particle?

In classical physics, we expect to know exactly where an object is at any given time. But quantum mechanics doesn't work that way. Instead, a particle's location is described as a **probability cloud**:

• The particle could be here, there, or somewhere in between.

• The probability of finding it in a particular spot depends on the intensity of its wave at that location.

This is where the quantum world gets strange: the particle doesn't have a single location until it's measured. Before that, it's as if it exists in all the possible locations at once, each with a certain likelihood.

• •

Relatable Example: Flipping a Coin

To make this idea more tangible, let's consider flipping a coin. Before it lands, there's a 50% chance it will be heads and a 50% chance it will be tails. While the coin is in the air, its outcome is uncertain—it exists in a state of possibility.

Now imagine the coin doesn't land until you look at it. The simple act of observing "collapses" the possibilities into a single reality: heads or tails. This is a simplified way to think about what happens to quantum particles when observed. The difference is that, unlike the coin, quantum particles don't have a fixed state until they're measured.

· ·

The Role of Uncertainty

One of the most famous principles in quantum mechanics is **Heisenberg's Uncertainty Principle**, which says we can't simultaneously know a particle's exact position and momentum (its speed and direction). The more precisely we measure one, the less precisely we can know the other.

This isn't due to a lack of technology or precision; it's a fundamental property of the quantum world. Imagine trying to measure the exact position of a ripple on a pond. The act of measuring it disturbs the ripple, making it impossible to get a perfect measurement of both its position and movement.

· ·

Quantum Superposition: Being in Two Places at Once

Another strange feature of quantum mechanics is **superposition**, the idea that particles can exist in multiple states simultaneously until observed. For example:

·   An electron can be in two places at once, traveling multiple paths simultaneously.

·   A photon of light can act as both a wave and a particle at the same time.

This defies our everyday logic, where things are either one way or another. But in the quantum world, possibilities coexist until something forces a decision.

· ·

Everyday Analogy: Quantum Choices and Coin Tosses

Think of superposition like standing at a crossroads. You could go left, right, or straight ahead, but until you start walking (or someone observes your choice), all these possibilities remain open. Once you commit to a direction, the potential paths "collapse" into the one you've chosen. Quantum particles live in this state of potentiality, and observation acts like the decision that determines their fate.

• •

How This Impacts the Observer Effect

So why does this matter for the observer effect? Here's the connection:

- Particles behave like waves when they're unobserved, existing in multiple possible states.

- Observation collapses the wave into a single state, forcing the particle to "choose" a definite location or behavior.

This means the act of observing isn't passive—it actively shapes what's being observed. In quantum mechanics, the observer isn't just a witness to reality but a participant in creating it.

• •

Summary

The world of quantum mechanics is a strange and beautiful place where particles act like waves, probabilities rule, and observation shapes reality. By understanding concepts like wave-particle duality, superposition, and the uncertainty principle, we begin to see how profoundly different the quantum world is from the one we experience every day.

In the next chapter, we'll explore the famous **double-slit experiment**, a groundbreaking demonstration of the observer effect that continues to puzzle and inspire scientists to this day. Through this experiment, we'll see just how powerful observation can be in shaping the behavior of particles—and perhaps, by extension, the world around us.

CHAPTER 3: THE DOUBLE-SLIT EXPERIMENT MADE SIMPLE

The double-slit experiment is one of the most famous and puzzling demonstrations in quantum mechanics. It shows how particles like electrons and photons behave differently depending on whether they're being observed. This experiment reveals the profound role of observation in shaping reality and serves as a cornerstone for understanding the observer effect.

In this chapter, we'll break down the double-slit experiment step by step, using analogies and thought experiments to make it approachable and relatable.

• •

STEP 1: SETTING THE STAGE

Imagine you have a barrier with two narrow slits cut into it, and behind this barrier is a blank screen. On one side of the barrier, you fire tiny particles—like electrons or photons—at the slits. The question is: what will appear on the screen?

• •

STEP 2: UNOBSERVED PARTICLES BEHAVE LIKE WAVES

If no one is observing the particles as they travel through the slits, something astonishing happens: the particles behave like waves. Instead of forming two neat lines on the screen (as you might expect if they traveled in straight lines through the slits), they create an **interference pattern**. This pattern looks like a series of light and dark bands, similar to ripples overlapping in a pond.

- **What's happening?**

  - Each particle acts like a wave, passing through **both slits simultaneously** and interfering with itself on the other side.

  - This interference causes the particles to land in areas of constructive interference (light bands) or destructive interference (dark bands).

- **Everyday comparison**: Picture sunlight streaming through window blinds. As the light passes through the slats, it creates overlapping patterns on the floor. These patterns are the result of wave interference.

• •

STEP 3: OBSERVING PARTICLES CHANGES THEIR BEHAVIOR

Now let's introduce an observer. Suppose you place a detector at the slits to see which slit each particle goes through. The result? The interference pattern disappears, and the particles behave like tiny marbles, creating two straight lines on the screen that correspond to the slits.

- **What's happening?**

   - When the particles are observed, they stop acting like waves and start acting like particles.

   - The mere act of measuring their path forces the particles to "choose" one slit or the other, collapsing the wave-like behavior into a single, definite outcome.

This astonishing shift highlights the observer effect: the act of observing changes the nature of reality.

• •

STEP 4: THE MYSTERY DEEPENS

What's truly mind-boggling is that this effect occurs even if the observation is indirect. For example:

- If you use a camera to record which slit the particles pass through but never look at the footage, the particles still behave like particles, not waves.

- It's as if the particles "know" they're being observed, even if no one ever sees the observation.

This raises profound questions: how does the act of observation influence behavior? What role does consciousness play in shaping reality?

• •

Using Visual Aids

To make this clearer, let's visualize the experiment step by step:

1. **Unobserved Particles**:

   - Imagine ripples in a pond spreading out and overlapping, creating areas where the waves amplify or cancel each other. This is similar to how particles behave like waves and create an interference pattern.

2. **Observed Particles**:

   - Now picture marbles rolling through two gates and landing in two neat piles. When particles are observed, they lose their wave-like nature and behave like marbles.

• •

A Thought Experiment: Collapsing Possibilities with Attention

To understand this concept on a more personal level, imagine this scenario: you're sitting in a room with a closed door. You know that behind the door, multiple possible outcomes could exist. For example:

- Someone could be standing outside the door.

- The hallway could be empty.

- A cat might be wandering by.

Until you open the door and observe what's on the other side, all these possibilities coexist. The moment you open the door, however, you "collapse" the possibilities into a single reality: you see what's actually there.

Now extend this idea to the quantum world. Particles exist in a haze of possibilities—a superposition—until someone observes them. Observation acts like opening the door, forcing the universe to "choose" an outcome.

• •

The Implications of the Double-Slit Experiment

The double-slit experiment isn't just a scientific curiosity; it has profound implications for how we understand the universe:

1. **Reality Is Influenced by Observation**:

   – In the quantum world, particles don't settle into a definite state until they're observed.

2. **The Observer Is Part of the System**:

   – Observation isn't a passive act—it actively shapes the outcome.

3. **Reality at Large**:

   – While the double-slit experiment applies to subatomic particles, it invites us to consider whether the observer effect could have broader implications for how we interact with and influence the world.

• •

Summary

The double-slit experiment reveals the strange and profound nature of quantum mechanics:

•   When unobserved, particles behave like waves, creating an interference pattern.

•   When observed, they behave like particles, choosing a definite path.

•   This shift demonstrates the observer effect in action: observation changes outcomes.

This experiment challenges our understanding of reality and forces us to consider the role of observation in shaping the universe. In the next chapter, we'll explore how this principle connects to daily life and how your attention might shape the outcomes you experience in your own world.

CHAPTER 4: FROM SCIENCE TO DAILY LIFE

Quantum mechanics may feel like it belongs in the realm of physicists and laboratories, but the concepts it reveals have surprising relevance to everyday life. The observer effect, in particular, isn't limited to subatomic particles—it has profound implications for how we live, make decisions, and shape our experiences.

In this chapter, we'll explore how attention and focus act as "observers" in your daily life, influencing the outcomes of everything from your health to your relationships. By the end, you'll see how the principles of quantum mechanics connect to tangible, practical changes in the world around you.

• •

The Observer Effect Beyond the Quantum Realm

In the quantum world, observation collapses possibilities into reality. At the human scale, attention and focus function similarly: what you observe—what you pay attention to— becomes your reality. This isn't mystical thinking; it's a psychological and behavioral truth. When you focus on something, you naturally align your actions and decisions to influence the outcome.

Let's take a simple example: imagine you're keeping a plant alive. If you regularly check the plant's soil moisture and adjust its watering schedule, the plant thrives. Your attention, much like the quantum observer, shapes the outcome.

This connection between observation and outcome has far-reaching implications for everyday life.

• •

How Attention and Focus Act as "Observers"

In daily life, your focus determines where your energy flows. When you observe or track something, you influence it simply by being aware of it. Here are two key ways this happens:

1. **Observation Brings Awareness**:

   –   When you monitor a habit or behavior, you gain insight into patterns you might otherwise ignore. This awareness helps you make conscious changes.

   –   Example: Keeping a food diary makes you more mindful of what you eat, often leading to healthier choices.

2. **Observation Drives Change**:

   –   Actively observing something creates accountability and motivation. Knowing that you're tracking progress often pushes you to perform better.

   –   Example: Tracking your steps with a fitness app often leads to increased activity because you're more aware of how much (or how little) you're moving.

• •

Practical Examples: Shaping Your Reality Through Observation

Example 1: Monitoring Your Diet

Imagine you've decided to eat healthier. Without observation, it's easy to slip into old habits—grabbing an extra cookie here or skipping a vegetable there. But when you start tracking your meals, something shifts:

•   You become aware of what you're eating and how it aligns with your goals.

•   The act of recording creates a sense of accountability, making you less likely to overindulge.

•   Over time, you notice patterns (e.g., you snack more when stressed) and can adjust your behavior accordingly.

THE OBSERVER'S REALITY: UNLOCKING THE HIDDEN POWER OF YOUR ATTE...

This simple act of observing your diet—writing down what you eat—collapses the possibilities of mindless eating into the reality of healthier habits.

• •

Example 2: Tracking Finances

Consider another example: managing your finances. Many people feel overwhelmed by money because they avoid looking at their accounts. But when you start tracking your spending, the observer effect comes into play:

- You become more mindful of where your money is going, identifying wasteful habits.

- You adjust your behavior, perhaps by cutting back on unnecessary expenses.

- Over time, the simple act of observing your finances can lead to significant savings and a sense of control.

In this case, observation turns financial chaos into financial clarity.

• •

Example 3: Observing Relationships

Observation isn't just about tracking numbers; it's also about paying attention to the people in your life. Relationships thrive when you notice and acknowledge others' needs and emotions:

- By observing how your partner reacts to stress, you can adjust your approach to be more supportive.

- By paying attention to small acts of kindness, you reinforce positive behaviors and deepen your bond.

In this way, the observer effect applies to the social world, where attention and focus shape the quality of your relationships.

• •

The Science of Attention: Why Observation Works

Why does observation have such a powerful effect on outcomes? The answer lies in both psychology and neuroscience:

1. **The Brain's Focus Mechanism**:

   – Your brain filters out unimportant information and prioritizes what you consciously observe. This is known as the **reticular activating system (RAS)**. For example, if you're tracking your diet, your brain starts to notice opportunities for healthier eating.

2. **Feedback Loops**:

   – Observation creates a feedback loop. When you see progress—like losing weight or saving money—it motivates you to continue the behavior, reinforcing the positive cycle.

3. **Mindfulness and Self-Awareness**:

   – Observation requires mindfulness, which is the act of being present and aware. Studies show that mindfulness improves focus, reduces stress, and enhances decision-making.

• •

A Thought Experiment: Observing Your Own Life

To connect the dots, try this thought experiment:

1. Choose one area of your life you'd like to improve—your health, finances, or even happiness.

2. Spend a week simply observing it. For example:

   – Write down everything you eat.

   – Record your daily spending.

   – Reflect on moments that made you happy or stressed.

By the end of the week, you'll likely notice patterns you hadn't seen before. This awareness is the first step toward change, as observation collapses vague possibilities into clear action steps.

• •

Bringing It All Together

The observer effect teaches us that observation isn't passive—it's a powerful tool for shaping reality. In the quantum world, observation changes particles. In everyday life, attention and focus change habits, relationships, and outcomes.

By observing the areas of your life you care about, you can begin to steer them in the direction you want. Whether it's tracking your diet, monitoring your finances, or paying closer attention to loved ones, observation empowers you to turn potential into reality.

In the next chapter, we'll explore how the observer effect relates to psychology and belief systems, diving into phenomena like the placebo effect and self-fulfilling prophecies. You'll see how even your expectations can shape your experience of the world.

• •

CHAPTER 5: ATTENTION SHAPES REALITY

What you focus on grows. This simple yet profound idea lies at the heart of the psychological observer effect. In quantum mechanics, particles respond to observation by settling into a specific state. Similarly, in your personal life, your attention acts like a spotlight, amplifying whatever it shines on. Whether it's your strengths, challenges, or aspirations, your focus shapes how these aspects of your life evolve.

In this chapter, we'll explore how attention influences your inner and outer world, backed by psychological principles and relatable examples. By the end, you'll have a practical mindfulness exercise to start harnessing the power of your focus to shape your reality.

• •

The Psychological Observer Effect: What You Focus On Grows

The psychological observer effect is the idea that attention magnifies the things we focus on. Just like observing a particle in quantum mechanics collapses its possibilities, focusing on something in life gives it more weight, clarity, and energy.

1. **The Brain's Spotlight: Reticular Activating System (RAS)**

    – The RAS is a network in your brainstem that acts as a filter, determining what information gets your conscious attention.

    – Example: If you buy a red car, you suddenly notice red cars everywhere. Your RAS highlights red cars because your attention has made them relevant.

2. **Self-Fulfilling Focus**:

–   If you focus on problems, they seem to grow, dominating your thoughts and perceptions.

–   If you focus on opportunities, you're more likely to see and seize them.

• •

How Focusing on Strengths Builds Confidence

Imagine two students preparing for an exam. One focuses on their weaknesses, worrying about what they don't know. The other focuses on their strengths, reviewing what they've mastered and building on it. The first student feels overwhelmed, while the second gains confidence and momentum. Why?

•   **Focus and Self-Perception**:

–   When you focus on your strengths, you reinforce a positive self-image. You start to see yourself as capable and resourceful, which boosts your confidence.

–   This doesn't mean ignoring weaknesses—it means approaching them with the mindset that your strengths can help you overcome challenges.

•   **Practical Example**:

–   Think of a public speaker who acknowledges their strength in storytelling but struggles with nervousness. By focusing on their strength, they're more likely to captivate their audience and feel empowered, rather than fixating on their fear.

• •

Examples of Attention Shaping Reality

Example 1: Relationships

In relationships, where you focus your attention determines how you perceive and interact with others:

- Focusing on someone's flaws can lead to resentment and distance.

- Focusing on their positive qualities can strengthen the bond and improve the relationship.

Example 2: Career

In the workplace, employees who focus on their achievements tend to feel more fulfilled and motivated, while those who dwell on their shortcomings often feel stuck and discouraged.

Example 3: Physical Health

Athletes often use mental imagery to enhance performance. By focusing on success—visualizing themselves crossing the finish line—they increase their confidence and drive, which translates into better physical results.

• •

The Science Behind Focus

1. **Neuroplasticity**:

   – Your brain rewires itself based on where you direct your attention. This process, called neuroplasticity, means that focusing on positive thoughts and actions can create stronger neural pathways for positivity over time.

2. **The Placebo Effect**:

   – Belief and attention can influence physical outcomes. For example, patients who believe a treatment will work often experience real improvements, even if the treatment is a placebo. This is attention shaping reality in action.

3. **Energy Follows Focus**:

   – The more attention you give something, the more mental and emotional energy you allocate to it. This is why focusing on gratitude can increase happiness, while focusing on grievances can amplify dissatisfaction.

• •

Practical Activity: Observing Your Thoughts

To harness the psychological observer effect, you need to become aware of where your attention naturally goes. This mindfulness exercise will help you observe your thoughts and their effects on your emotions and actions.

STEP 1: FIND A QUIET SPACE

- Set aside 5–10 minutes in a quiet, comfortable place where you won't be interrupted.

STEP 2: CLOSE YOUR EYES AND BREATHE

- Take a few deep breaths to center yourself. Focus on the sensation of your breath moving in and out.

STEP 3: OBSERVE YOUR THOUGHTS

- Let your mind wander, and simply notice the thoughts that arise. Avoid judgment—don't label thoughts as good or bad.

- Ask yourself: *What am I focusing on?* Is it positive, negative, or neutral?

STEP 4: NOTICE THE EFFECTS

- Pay attention to how each thought makes you feel. Does it uplift or drain you? Does it spark action or hesitation?

STEP 5: REDIRECT YOUR FOCUS

- If you notice a negative or unhelpful thought, consciously redirect your attention to something positive or constructive. For example:

    – Replace "I'm terrible at this" with "I'm learning and improving."

    – Replace "I'll never achieve this" with "What's one small step I can take?"

STEP 6: REFLECT

•   After the exercise, jot down your observations. What patterns did you notice? How did redirecting your focus change your feelings or perspective?

• •

Daily Applications of Focus

1. **Set a Positive Intention Each Morning**:

   –   Start your day by identifying one thing you want to focus on. For example, "Today, I'll focus on gratitude," or "I'll focus on giving my best effort."

2. **Track Your Attention Triggers**:

   –   Throughout the day, notice what grabs your attention. Is it productive or draining? Redirect as needed.

3. **Practice Gratitude**:

   –   Spend a few moments each day focusing on things you're grateful for. This simple act can shift your mindset and improve your overall well-being.

• •

Summary

Attention shapes reality in profound ways. Like the quantum observer effect, your focus determines what grows and thrives in your life. Whether it's building confidence by focusing on your strengths, improving relationships by noticing positive qualities, or enhancing performance through visualization, the power of attention is transformative.

By practicing mindfulness and intentionally directing your focus, you can create a reality that aligns with your goals and values. In the next chapter, we'll explore how expectations—what you believe will happen—can shape your experiences through phenomena like the placebo effect and self-fulfilling prophecies.

CHAPTER 6: THE PLACEBO EFFECT

The placebo effect is one of the most fascinating examples of how belief shapes reality. In medicine, it's often seen as a quirk —a phenomenon where patients experience real improvements from treatments that have no active ingredients. However, the implications of the placebo effect go far beyond medicine, highlighting the power of belief and attention to influence outcomes in unexpected ways.

In this chapter, we'll explore the science behind the placebo effect, dive into real-life case studies, and draw connections between this phenomenon and quantum principles. Finally, we'll guide you through an activity to explore the power of belief in your own life.

• •

What Is the Placebo Effect?

The placebo effect occurs when a person experiences real physical or emotional improvement after receiving a treatment that has no therapeutic effect. For example:

- A patient given a sugar pill instead of an actual medication may still experience pain relief or other positive outcomes, simply because they believe the treatment will work.

This effect demonstrates the incredible influence of belief on the body and mind. While it's often studied in the context of medical treatments, the placebo effect reveals a universal truth: **what we believe can shape our reality.**

• •

How Belief Influences Outcomes

1. **The Brain's Role**:

-    When you believe something will help, your brain activates pathways associated with healing and relief. For example:

    •   Believing in a painkiller triggers the release of endorphins, the body's natural painkillers.

    •   Expecting recovery can boost immune system responses.

2. **The Power of Expectation**:

-    The placebo effect is closely tied to expectation. When you expect a positive outcome, your body and mind align to make that outcome more likely.

3. **Perception and Reality**:

-    Belief doesn't just influence how we feel—it can change what we perceive. A placebo treatment can make pain feel less intense or symptoms feel less severe, even though nothing physical has changed in the treatment itself.

• •

Real-Life Case Studies of the Placebo Effect

Case Study 1: Sham Surgeries

In a groundbreaking study, patients with knee pain were divided into two groups:

•   One group received actual knee surgery.

•   The other group received a "sham" surgery, where doctors made incisions but didn't perform the procedure.

Amazingly, both groups reported significant pain relief. The belief that they had undergone surgery was enough to trigger real improvement.

• •

Case Study 2: Fake Pills, Real Relief

In a study on migraine treatments:

- One group received a real migraine medication.

- Another group received a placebo pill labeled as a placebo. Even when patients knew they were taking a placebo, they still experienced reduced pain. This shows that belief can work even when we're aware of its mechanisms.

• •

Case Study 3: Placebo Injections

In a study on Parkinson's disease, patients received saline injections instead of actual dopamine-enhancing medication. Despite this, their symptoms improved. Belief in the treatment activated dopamine production in the brain, mimicking the effects of real medication.

• •

Connection to Quantum Principles

The placebo effect mirrors quantum principles in surprising ways:

1. **The Observer Effect**:

   – Just as observing a quantum particle collapses its possibilities into a single reality, belief focuses attention and expectation, shaping the body's response.

   – In both cases, unseen forces—attention and belief—play an active role in determining outcomes.

2. **Probabilities and Possibilities**:

   – The quantum world is governed by probabilities, and observation determines the outcome. Similarly, the placebo effect demonstrates that belief tilts the probabilities toward healing or improvement.

3. **Mind-Body Connection**:

   – Quantum mechanics blurs the line between observer and observed. The placebo effect similarly blurs the line between mind and body, showing how belief can manifest as physical change.

• •

Activity: Harnessing the Placebo Effect in Your Life

To explore the power of belief in your own life, try this simple activity:

STEP 1: IDENTIFY AN OUTCOME YOU WANT

- Choose a specific area of your life where you'd like to see improvement. For example:

  – Reducing stress.

  – Improving your focus.

  – Enhancing physical recovery from an injury.

STEP 2: WRITE DOWN YOUR BELIEF

- Write a positive, believable statement about the outcome you want. For example:

  - "I believe I will feel calmer and more focused this week."

  - "I believe my body is healing, and I will feel stronger each day."

STEP 3: VISUALIZE THE OUTCOME

- Spend a few moments each day imagining the desired outcome as if it has already happened. Picture yourself calm, focused, or physically strong.

STEP 4: TRACK YOUR PROGRESS

- Over the next week, observe how your belief influences your thoughts, actions, and feelings. Jot down any changes you notice, however small.

STEP 5: REFLECT

- At the end of the week, review your observations. Did your belief influence your actions or mindset? Did you notice any positive changes?

This activity helps you experience firsthand how focusing your belief can shape your reality, much like the placebo effect in action.

• •

The Placebo Effect and Everyday Life

The placebo effect isn't limited to medicine—it shows up in daily life in subtle ways:

- **Confidence Boosters**: Wearing an outfit you believe makes you look good can change how you feel and act.

- **Performance Enhancers**: Believing in your abilities can improve your performance, whether in sports, work, or creative endeavors.

- **Health Habits**: Believing in the benefits of a healthy routine (like exercise or meditation) makes you more likely to stick with it, enhancing its effectiveness.

• •

Summary

The placebo effect demonstrates the incredible power of belief to shape reality, even in measurable physical outcomes. From sham surgeries to placebo pills, real-life case studies show that expectation and focus can drive remarkable changes. By understanding the placebo effect, we see parallels to quantum principles and gain a deeper appreciation for the role of unseen forces like attention and belief.

By practicing the activity outlined above, you can begin to harness the placebo effect in your own life, using the power of belief to create positive changes. In the next chapter, we'll explore another fascinating way belief shapes reality: **self-fulfilling prophecies**, where our expectations influence outcomes in surprising and profound ways.

CHAPTER 7: SELF-FULFILLING PROPHECIES

A self-fulfilling prophecy is the idea that what we expect to happen often becomes reality, not because of magic or coincidence, but because our expectations shape our behavior, decisions, and interactions. It's as if our beliefs about ourselves and the world act like invisible scripts, guiding how events unfold. In this chapter, we'll delve into the science of self-fulfilling prophecies, explore real-world examples in parenting, education, and personal goals, and provide a practical exercise to reframe limiting beliefs into empowering expectations.

• •

What Is a Self-Fulfilling Prophecy?

A self-fulfilling prophecy occurs when:

1. You hold a belief or expectation about a person, situation, or outcome.

2. That belief influences your behavior in ways that make the expected outcome more likely to occur.

3. The outcome reinforces your original belief.

This cycle can be positive or negative:

- A positive belief can lead to success and confidence.

- A negative belief can reinforce failure and self-doubt.

• •

How Expectations Influence Behavior and Results

Expectations shape behavior in subtle but powerful ways:

1. **Your Actions Align with Expectations**:

   - If you expect to succeed, you're more likely to take steps that lead to success, like preparation and persistence.

   - If you expect to fail, you might give up early or avoid challenges altogether.

2. **Your Perception Changes**:

   - Expectations act as filters, shaping how you interpret events.

   - For example, if you expect someone to be rude, you might interpret neutral behavior as negative, reinforcing your belief.

3. **Other People Respond to Your Beliefs**:

   - Your expectations influence how you treat others, which affects how they respond to you. This creates a feedback loop that can reinforce your belief.

• •

Real-Life Examples of Self-Fulfilling Prophecies

Example 1: Parenting

Parental expectations often have a profound impact on children's development:

- **Positive Expectations**:

  - Parents who believe their child is capable and intelligent tend to encourage exploration, provide opportunities for growth, and celebrate achievements. This fosters confidence and a growth mindset in the child.

- **Negative Expectations**:

  - Parents who expect their child to struggle or misbehave may unconsciously communicate disappointment or frustration, which can undermine the child's self-esteem and performance.

• •

Example 2: Education

Teachers' expectations can significantly influence students' academic performance:

- **The Pygmalion Effect**:

  - In a famous study, researchers told teachers that certain students were expected to excel academically, based on test results (which were randomly assigned). Over time, these students performed better, not because they were inherently more capable, but because teachers' expectations shaped how they interacted with them.

- **The Opposite Effect**:

  - Low expectations can result in less encouragement, fewer opportunities, and diminished student outcomes, reinforcing the belief that the student is incapable.

• •

Example 3: Personal Goals

Your own expectations play a key role in achieving personal goals:

- **Positive Beliefs**:

  - If you believe you can complete a marathon, you're more likely to train consistently, overcome setbacks, and succeed.

- **Limiting Beliefs**:

  - If you believe you're "not good with money," you might avoid budgeting or investing, reinforcing financial struggles.

• •

The Science Behind Self-Fulfilling Prophecies

1. **Cognitive Biases**:

– Expectations create a lens through which we see the world. Confirmation bias leads us to seek out and remember information that supports our beliefs, reinforcing them over time.

2. **Neuroplasticity**:

– Repeated thoughts and beliefs strengthen neural pathways in the brain. Positive beliefs reinforce pathways that support confidence and resilience, while negative beliefs reinforce fear and doubt.

3. **The Role of Subconscious Behavior**:

– Much of the behavior influenced by self-fulfilling prophecies happens subconsciously. For instance, your tone of voice, body language, and word choice can align with your expectations without you realizing it.

• •

Exercise: Reframing Limiting Beliefs
The key to breaking negative self-fulfilling prophecies is to identify limiting beliefs and reframe them into positive, empowering expectations. This exercise will guide you through the process.

• •

STEP 1: IDENTIFY A LIMITING BELIEF

- Think of an area in your life where you feel stuck or struggle to achieve your goals. Ask yourself:

  - What belief might be holding me back?

  - Examples:

    - "I'm not good enough to get that promotion."

    - "I always fail at relationships."

    - "I'm terrible with money."

· ·

STEP 2: ANALYZE THE BELIEF

- Reflect on how this belief influences your behavior:

  – Does it make you avoid opportunities or take fewer risks?

  – Does it reinforce fear or self-doubt?

• •

STEP 3: REFRAME
THE BELIEF

- Replace the limiting belief with a positive, realistic expectation. Use empowering language that aligns with your goals.

  - Examples:

    - "I am learning and growing, and I'm ready for new challenges."

    - "I deserve healthy, fulfilling relationships, and I'm taking steps to build them."

    - "I can improve my financial skills and make better choices."

• •

STEP 4: TAKE ACTION

- Identify one small action that aligns with your new belief.

  - Examples:

    - Apply for a job you've been hesitant about.

    - Have an honest conversation with a loved one about improving your relationship.

    - Create a simple budget to start managing your money.

• •

STEP 5: REFLECT AND REINFORCE

- Over time, observe how this new belief influences your behavior and outcomes. Celebrate small successes to reinforce the positive cycle.

• •

Practical Application in Daily Life

Self-fulfilling prophecies don't just shape big events; they influence everyday interactions and habits:

1. **Set Positive Expectations**:

   – Start each day with a positive intention, such as, "Today, I will focus on what I can control and do my best."

2. **Reframe Negative Thoughts**:

   – When you catch yourself thinking, "This will never work," pause and reframe it as, "What if this works?"

3. **Acknowledge Progress**:

   – Recognize and celebrate small victories. This reinforces the belief that progress is possible.

• •

Summary

Self-fulfilling prophecies show how our expectations shape behavior and results. Whether in parenting, education, or personal goals, beliefs act like invisible scripts, guiding how events unfold. By reframing limiting beliefs into empowering expectations, you can harness the psychological observer effect to create positive outcomes.

In the next chapter, we'll explore the fascinating world of habits and patterns, showing how observing your behaviors can help you break negative cycles and build a life aligned with your goals.

CHAPTER 8: OBSERVING YOUR HABITS

Habits shape much of our daily lives. From how we start our mornings to how we wind down at night, these automatic behaviors create the foundation for success or struggle. Yet, many of us remain unaware of the habits that guide our actions. This lack of awareness can trap us in cycles we wish to break—whether it's procrastination, overeating, or negative self-talk.

In this chapter, we'll explore how observing your habits is the first step toward meaningful change. You'll learn about the "habit loop," how observation disrupts negative patterns, and how to create space for positive behaviors. Finally, a practical journaling activity will help you shine a light on your habits and start reshaping them to align with your goals.

• •

The Power of Observation

Observation is more than just noticing—it's an intentional act of awareness. When you observe your habits, you bring unconscious patterns into the conscious mind. This awareness is transformative because:

1. **You Gain Clarity**:

   – By observing your actions, you see what's working for you and what isn't. Awareness provides the foundation for making deliberate changes.

2. **You Interrupt Automatic Behavior**:

–   Observation creates a moment of pause. It allows you to stop running on autopilot and consider alternative actions.

3. **You Empower Change**:

–   Once you're aware of a habit, you can analyze it, reframe it, and replace it with something more constructive.

• •

The Habit Loop: Understanding Your Behaviors

Every habit follows a predictable cycle called the **habit loop**. This loop consists of three parts:

1. **Cue**:

–   A trigger that prompts the habit. It could be a time of day, a specific emotion, or an environmental factor.

–   Example: Feeling bored might cue you to check your phone.

2. **Routine**:

–   The behavior itself—what you do in response to the cue.

–   Example: You scroll through social media for 20 minutes.

3. **Reward**:

–   The benefit or relief you experience from completing the routine. Rewards reinforce the habit, making it more likely to repeat.

–   Example: Checking your phone provides a momentary escape from boredom, reinforcing the behavior.

• •

Breaking Negative Patterns with Observation

To change a habit, you need to disrupt the habit loop. Observation is the key to this disruption:

1. **Identify the Cue**:

- Observation helps you recognize what triggers the habit. Is it an emotion, a time of day, or a specific situation?

- Example: You notice that you snack mindlessly whenever you feel stressed.

2. **Analyze the Routine**:

- Once you know the cue, you can observe the habitual behavior itself. What are you doing, and why?

- Example: You realize you're reaching for snacks because it's a quick way to distract yourself.

3. **Evaluate the Reward**:

- Finally, reflect on the reward you're seeking. Is it relief, comfort, or stimulation? Understanding the reward helps you find healthier alternatives.

- Example: You realize the reward is stress relief, and you can achieve it through exercise or deep breathing instead.

. . . .

How Observation Creates Space for Change

When you observe your habits, you create a moment of mindfulness between the cue and the routine. This pause allows you to:

• **Question Your Choices**:

- Instead of reacting automatically, you can ask, "Is this behavior serving me?"

• **Experiment with Alternatives**:

- Replace the routine with a healthier action that provides a similar reward.

- Example: Instead of snacking when stressed, you take a short walk or practice deep breathing.

. . . .

Practical Example: Observing and Changing a Habit

Let's say you want to reduce screen time before bed. By observing your habit loop, you might discover:

- **Cue**: Feeling tired but restless triggers the desire to check your phone.

- **Routine**: You scroll through social media for an hour.

- **Reward**: You feel entertained or distracted from end-of-day stress.

With this awareness, you can replace the routine:

- Set a specific bedtime and charge your phone outside the bedroom.

- Replace screen time with reading a book or practicing relaxation techniques.

- Over time, the new habit becomes as automatic as the old one.

• •

Activity: Journal Your Habits for a Week

This activity will help you observe your habits and gain the clarity needed to make intentional changes.

• •

STEP 1: CHOOSE AN AREA TO OBSERVE

- Focus on one category of habits, such as:
  - Morning routines.
  - Eating habits.
  - Work productivity.
  - Evening wind-down rituals.

· ·

STEP 2: KEEP A DAILY JOURNAL

- For one week, write down the following:
  - The time of day.
  - The specific habit you engaged in.
  - What triggered the habit (the cue).
  - How you felt before and after the habit (the reward).
- Example entry:
  - *Time*: 3:00 PM.
  - *Habit*: Grabbed a coffee and a snack.
  - *Cue*: Feeling fatigued after lunch.
  - *Reward*: Temporary energy boost.

• •

STEP 3: ANALYZE YOUR PATTERNS

•   At the end of the week, review your journal. Look for recurring cues, routines, and rewards. Ask yourself:

-     Which habits are serving me?

-     Which habits are holding me back?

-     What triggers my negative habits?

• •

STEP 4: CHOOSE ONE HABIT TO CHANGE

- Select a habit you want to improve. Use the habit loop to guide your change:

    - Identify the cue.

    - Replace the routine with a positive behavior.

    - Ensure the new routine provides a similar reward.

· ·

STEP 5: TRACK YOUR PROGRESS

- For the next week, track your efforts to replace the habit. Celebrate small victories and adjust as needed.

. . . .

The Ripple Effect of Observing Habits

Observing and changing one habit often leads to positive ripple effects:

- Improving your morning routine might increase productivity throughout the day.

- Reducing screen time might improve sleep quality, which enhances overall health.

- Developing mindfulness around eating habits might lead to better physical and emotional well-being.

By observing your habits, you take control of the small, daily behaviors that shape your life. Over time, this awareness leads to lasting change and a life aligned with your values and goals.

. . . .

Summary

Habits are the building blocks of your life, and observing them is the first step toward transformation. By understanding the habit loop—cue, routine, and reward—you can identify triggers, disrupt negative patterns, and create space for positive change. The journaling activity provides a practical way to bring awareness to your habits, empowering you to make intentional choices.

In the next chapter, we'll explore how the observer effect extends to relationships and social dynamics, shaping how we interact with others and how they respond to us.

CHAPTER 9: CONSCIOUSNESS AS THE ULTIMATE OBSERVER

In the quantum world, the observer plays a pivotal role in shaping reality. Particles exist as probabilities until they are observed, at which point they "collapse" into a specific state. This phenomenon has led to one of the most profound questions in science and philosophy: what is the ultimate observer? Could consciousness itself be the force that defines reality?

In this chapter, we'll explore theories of consciousness and its relationship to the quantum world. We'll examine how awareness shapes reality and dive into the deep philosophical implications of the observer effect, asking: *Does the observer create the universe?*

• •

Consciousness and the Quantum World

Consciousness has been described as one of the great mysteries of existence. While scientists understand many aspects of how the brain works, the essence of subjective experience—what it feels like to be you—remains elusive. The observer effect in quantum mechanics adds another layer of mystery, suggesting that consciousness might play a fundamental role in shaping the fabric of reality.

1. **Quantum Measurement and Observation**:

   – In quantum mechanics, particles exist in a superposition

of states until measured. This raises the question: what constitutes a "measurement"? Is it the act of a human observer, the interaction with a measuring device, or something else entirely?

–   Some interpretations, like the **Copenhagen Interpretation**, suggest that observation by a conscious observer is crucial to collapsing the quantum wave function into a definite reality.

2. **The Mind-Matter Connection**:

–   If consciousness influences the quantum world, it implies a profound connection between the mind (subjective experience) and matter (objective reality).

• •
Theories Linking Consciousness and Quantum Mechanics
1. **The Von Neumann-Wigner Interpretation**:

–   Physicist John von Neumann and mathematician Eugene Wigner proposed that consciousness is essential for the collapse of the wave function. According to this interpretation, reality remains in a state of potentiality until observed by a conscious mind.

2. **Penrose-Hameroff Orch-OR Theory**:

–   Physicist Roger Penrose and anesthesiologist Stuart Hameroff suggest that consciousness arises from quantum processes within the brain's microtubules (tiny structures in neurons). This theory posits that consciousness itself operates at the quantum level, potentially influencing the physical world.

3. **The Participatory Universe (John Wheeler)**:

–   Physicist John Wheeler proposed that the universe is fundamentally participatory. Observers don't just passively witness reality—they actively shape it. Wheeler's famous phrase, "No phenomenon is a phenomenon until it is an observed phenomenon," underscores the idea that observation is essential to existence.

. . . .

Awareness and the Creation of Reality

The role of awareness in shaping reality extends beyond quantum mechanics. On a practical level, our perceptions and focus determine how we experience and interact with the world. Philosophers and scientists have long debated whether consciousness merely observes reality or actively creates it.

1. **Subjective Reality**:

 – Each person's experience of the world is shaped by their awareness, beliefs, and perceptions. Two people can observe the same event but interpret it differently based on their individual consciousness.

2. **The Power of Focus**:

 – Attention acts like a filter, selecting which aspects of reality to prioritize. This idea aligns with the psychological observer effect, where focusing on specific elements (e.g., strengths, challenges, opportunities) amplifies their presence in your life.

. . . .

Philosophical Questions: Does the Observer Create the Universe?

1. **Reality as a Construct**:

 – If observation determines reality at the quantum level, could it be true on a larger scale? Is the universe a construct of collective consciousness, shaped by the sum of all observers?

2. **The Chicken-and-Egg Problem**:

 – If the universe requires an observer to exist, who or

what was the first observer? This question has profound implications for our understanding of existence and origins.

3. **Consciousness as the Fundamental Element**:

– Some theories suggest that consciousness is not a byproduct of matter but the foundation of reality itself. From this perspective, physical reality emerges from a universal consciousness rather than the other way around.

· ·

A Thought Experiment: The Tree in the Forest

Consider the classic philosophical question: *If a tree falls in a forest and no one is around to hear it, does it make a sound?* At first, this might seem trivial, but it raises deep questions about the nature of observation and existence:

- In the quantum framework, if no one observes the event, does the tree's fall exist in a definite state, or does it remain a cloud of probabilities?

- If consciousness is necessary to create reality, does unobserved reality exist at all?

· ·

Scientific Skepticism and Open Questions

While the connection between consciousness and quantum mechanics is intriguing, it remains a topic of debate among scientists. Some argue that:

1. The observer effect doesn't require human consciousness, as any interaction with a measuring device can collapse the wave function.

2. Consciousness might emerge from physical processes, rather than being a fundamental force shaping reality.

Despite these objections, the interplay between consciousness and the quantum world continues to inspire exploration, challenging the boundaries of what we know.

• •

Practical Implications: Conscious Creation

Even without definitive answers, the idea that awareness shapes reality has practical applications:

1. **Mindfulness and Intentionality**:

   – 	Practicing mindfulness trains you to observe your thoughts and actions, influencing your behavior and outcomes in positive ways.

2. **Visualization and Focus**:

   – 	Visualization techniques, often used by athletes and performers, rely on the principle that focused awareness can shape reality. By imagining success, you align your actions with your goals.

3. **Collective Consciousness**:

   – 	If consciousness shapes reality, collective awareness might influence larger societal outcomes. This raises the possibility that global focus and intention could drive positive change.

• •

Summary

Consciousness may be the ultimate observer, shaping reality at both the quantum and practical levels. Theories like the Von Neumann-Wigner interpretation, the participatory universe, and the Orch-OR model suggest that awareness plays a central role in collapsing possibilities into definite outcomes. While science has yet to fully unravel the mysteries of consciousness, its connection to the quantum world raises profound questions about the nature of existence.

Does the observer create the universe? While we may not have a definitive answer, one thing is clear: your consciousness has the power to shape your reality. In the next chapter, we'll explore how ancient wisdom aligns with quantum principles, revealing timeless insights about the connection between observation, reality, and the cosmos.

CHAPTER 10: "AS ABOVE, SO BELOW" – ANCIENT WISDOM MEETS QUANTUM PHYSICS

Throughout history, spiritual traditions and ancient teachings have offered profound insights into the nature of reality. These ideas, often seen as mystical or metaphorical, find striking parallels in the discoveries of quantum physics. One of the most resonant examples is the **Hermetic principle**: *"As above, so below; as within, so without."* This timeless phrase encapsulates the idea that the patterns governing the universe at the grandest scales mirror those found in the smallest details, and that inner states influence the outer world.

In this chapter, we'll explore how ancient wisdom aligns with quantum mechanics, focusing on the Hermetic principle and examples from traditions like Hinduism, Taoism, and others. These connections reveal a shared understanding of the interconnectedness of all things, suggesting that science and spirituality may be speaking different languages to describe the same truths.

• •

The Hermetic Principle: "As Above, So Below"

The phrase *"As above, so below; as within, so without"* originates from Hermeticism, an ancient philosophical system attributed to Hermes Trismegistus. This principle suggests:

1. **Cosmic Reflection**:

    – The macrocosm (the universe) reflects the microcosm (individual existence), and vice versa.

2. **Interconnectedness**:

    – The laws that govern the heavens also govern the earth, implying a unity in the fabric of reality.

3. **Inner-Outer Alignment**:

    – Inner states of mind, emotion, and spirit influence external circumstances.

This idea finds a modern echo in quantum physics, where patterns observed at the quantum level reflect broader truths about the universe.

• •

Quantum Mechanics and "As Above, So Below"

1. **Fractal Nature of Reality**:

    – Quantum physics reveals that reality is self-similar across scales. For example:

        • Subatomic particles exhibit behaviors that resemble larger systems, like galaxies orbiting a center.

        • Fractal mathematics, used to describe natural patterns (e.g., trees, rivers), applies equally to quantum systems and cosmic structures.

2. **Quantum Entanglement**:

    – When two particles become entangled, changes to one instantly affect the other, regardless of distance. This mirrors the Hermetic idea that all parts of the universe are intrinsically connected, no matter how far apart they appear.

3. **Observer Effect**:

   – The observer effect demonstrates that the act of observation shapes reality, aligning with the Hermetic notion that what happens "within" (the mind's focus and awareness) influences what occurs "without" (the external world).

● ●

Ancient Traditions and Quantum Parallels

1. Hinduism: The Unified Field

In Hindu philosophy, the concept of **Brahman** represents the ultimate, unified reality that underlies everything in existence. This idea resonates with quantum physics:

• **Brahman as the Quantum Field**:

   – Brahman is described as infinite, formless, and the source of all creation. Similarly, the quantum field is an underlying energy field from which all particles and forces emerge.

• **Atman and Brahman**:

   – The relationship between the individual soul (*Atman*) and the universal reality (*Brahman*) reflects *"As above, so below"*: the microcosmic self mirrors the macrocosmic whole.

● ●

2. Taoism: The Flow of Qi and Quantum Uncertainty

Taoism emphasizes the interconnectedness of all things through the concept of **Tao** (the Way) and **Qi** (life force):

• **Tao as Quantum Harmony**:

   – The Tao is the unifying principle that governs the flow of the universe. Like quantum mechanics, it describes a dynamic, ever-changing reality where balance and flow are key.

- **Yin and Yang**:

    –   The interplay of opposites (Yin and Yang) reflects quantum dualities, such as wave-particle duality. Just as Yin and Yang are inseparable, particles and waves coexist in quantum systems.

· ·

3. Hermeticism: The Mind as Creator
Hermetic teachings explicitly connect consciousness to the creation of reality:
- **"The All is Mind"**:

    –   Hermeticism teaches that the universe is mental in nature, created and influenced by thought. This aligns with the quantum notion that observation (and perhaps consciousness) plays a role in shaping reality.

· ·

4. Other Traditions
- **Buddhism**:

    –   The Buddhist concept of **interdependent origination**—that all phenomena arise in dependence on other phenomena—parallels quantum entanglement and the interconnectedness of reality.

- **Native American Teachings**:

    –   Indigenous traditions often emphasize the unity of all life and the idea that every part of the natural world is interconnected, reflecting *"As above, so below."*

· ·

Philosophical Implications
1. **Unity in Diversity**:

    –   Ancient teachings and quantum mechanics suggest that diversity in the universe arises from a single, unified source. Whether we call it Brahman, Tao, or the quantum field, the message is the same: all things are connected.

2. **Interconnectedness Across Scales**:

–   The principle *"As above, so below"* implies that understanding the small gives insight into the large, and vice versa. Quantum physics demonstrates this with its ability to explain cosmic phenomena through subatomic principles.

3. **The Role of Awareness**:

–   Both ancient wisdom and quantum mechanics point to the importance of observation and awareness in shaping reality. This raises profound questions about our role as conscious beings in the universe.

• •

Practical Applications: Living "As Above, So Below"
The principle *"As above, so below"* isn't just theoretical—it has practical implications for daily life:

1. **Mindful Reflection**:

–   Your inner state reflects outward into your relationships and circumstances. Cultivating inner peace and clarity can lead to harmony in your external world.

2. **Aligning with Nature**:

–   Observing patterns in nature can offer guidance for personal growth. For example, seasons remind us of the cyclical nature of life, while fractals teach us about balance and repetition.

3. **Holistic Decision-Making**:

–   Recognizing interconnectedness encourages holistic thinking, where decisions consider the well-being of the whole system, not just individual parts.

• •

Summary

The Hermetic principle *"As above, so below"* offers a timeless lens through which to view the interconnectedness of all things. From the unified reality of Hinduism to the dynamic flow of Taoism, ancient teachings align with quantum mechanics in remarkable ways, suggesting that the universe operates as a harmonious whole.

By exploring these connections, we see that science and spirituality are not opposing forces but complementary approaches to understanding reality. In the next chapter, we'll dive deeper into how observation influences social dynamics, exploring how your attention and focus shape relationships and interactions.

CHAPTER 11: THE OBSERVER EFFECT IN SOCIAL DYNAMICS

The observer effect, as seen in quantum mechanics, isn't confined to particles—it has profound implications in social dynamics. Humans, like quantum particles, often change their behavior when they know they're being observed. Whether it's subtle shifts in posture, speech, or actions, awareness of being watched can transform how we act and interact.

In this chapter, we'll explore how the observer effect manifests in social situations, using examples from social media, surveillance, and public speaking. Finally, we'll offer practical tips on how to use observation intentionally to create positive social outcomes.

• •

Behavior Changes Under Observation

When people realize they are being observed, their behavior often changes. This phenomenon is well-documented in psychology and sociology and can manifest in various ways:

1. **Self-Consciousness**:

   – People become more aware of their actions, leading to either heightened performance or increased anxiety.

2. **Social Conformity**:

   – Observation can encourage people to align their behavior with societal norms or expectations, even if they might act differently in private.

3. **Impression Management**:

  – Knowing they're being observed, individuals may adjust their actions to create a desired image, such as appearing more confident, competent, or likable.

• •

Examples of the Observer Effect in Social Dynamics

1. Social Media

Social media platforms are a prime example of how observation shapes behavior:

• **Curated Personas**:

  – On platforms like Instagram or LinkedIn, people often present idealized versions of themselves, carefully curating content to gain approval or admiration.

• **Echo Chambers**:

  – The awareness of an audience can lead to conformity with the views of a specific group, reinforcing beliefs and discouraging dissent.

• **Performance Anxiety**:

  – The pressure of constant observation can lead to stress and anxiety, especially among younger users.

2. Surveillance

In settings where surveillance is prevalent, such as workplaces or public spaces, people alter their behavior to avoid negative consequences:

• **Workplace Monitoring**:

  – Employees aware of performance tracking may work harder or adhere more strictly to company policies.

• **Public Behavior**:

  – The presence of security cameras often discourages illegal or antisocial actions, though it can also create a sense of paranoia.

3. Public Speaking

Public speaking provides a clear illustration of the observer effect:

- Many people report feeling nervous or awkward when speaking in front of an audience, even if they are confident in private conversations.

- Awareness of being watched can enhance focus and performance for some, but it can also trigger self-doubt in others.

· ·

The Psychology Behind the Observer Effect in Social Dynamics

1. **Self-Awareness Theory**:

   – Observation increases self-awareness, prompting individuals to compare their actions with their internal standards or societal expectations.

   – This can lead to improvement (if the standards are achievable) or discomfort (if there's a perceived mismatch).

2. **The Hawthorne Effect**:

   – This phenomenon, observed in workplace studies, shows that people perform better when they know they are being studied, even if no actual changes are made to their environment.

3. **The Spotlight Effect**:

   – People often overestimate how much others notice or judge their behavior, leading to heightened self-consciousness.

· ·

Using the Observer Effect to Positively Influence Social Outcomes

The observer effect doesn't have to be a source of anxiety. When used intentionally, it can create opportunities for growth, connection, and influence. Here are practical tips for leveraging observation in social dynamics:

1. Be a Positive Observer

- **Active Listening**:
  - When interacting with others, give them your full attention. Eye contact, nodding, and affirmations like "I see" or "I understand" make people feel valued and encourage openness.

- **Positive Feedback**:
  - Acknowledge and reinforce constructive behavior. For example, praising someone's effort or creativity encourages them to continue those actions.

2. Practice Mindful Observation

- **Focus on Nonverbal Cues**:
  - Pay attention to body language, tone of voice, and facial expressions to gain deeper insights into others' emotions and intentions.

- **Observe Without Judgment**:
  - Practice observing without immediately forming conclusions. This creates a safe space for authentic interaction.

3. Create an Empowering Presence

- **Model Confidence**:
  - Your behavior influences others. By projecting calm and confidence, you set a tone that encourages similar responses.

- **Set Clear Expectations**:
  - In leadership or teamwork scenarios, make your expectations transparent. When people know they're being observed in a positive, supportive way, they're more likely to rise to the occasion.

4. Use Observation to Build Connection

- **Mirror Behavior**:

  – Subtle mirroring of others' posture or gestures can create a sense of rapport and understanding.

- **Show Empathy**:

  – Reflect back what others share with phrases like "It sounds like you're feeling..." or "I understand how important this is to you."

• •

Ethical Considerations in Observation

While observation can be a powerful tool, it's essential to use it ethically:

1. **Respect Privacy**:

   – Avoid observing or monitoring others in ways that invade their personal space or boundaries.

2. **Be Transparent**:

   – In professional settings, ensure people are aware of observation practices and their purpose.

3. **Avoid Manipulation**:

   – Use observation to empower and connect, not to exploit or control others.

• •

The Broader Implications of Social Observation

The observer effect in social dynamics raises important questions about society and technology:

1. **Surveillance Culture**:

   – As observation becomes increasingly automated (e.g., through AI and facial recognition), how does it shape human behavior and freedom?

2. **The Role of Social Media**:

–   Platforms amplify the observer effect by creating a virtual audience for every action. What does this mean for authenticity and mental health?

3. **Collective Observation**:

–   Just as collective focus in quantum systems can shape reality, shared attention in social movements can drive cultural and political change.

· ·

Summary

The observer effect in social dynamics shows that people change their behavior when they know they're being observed. From social media to public speaking, observation shapes actions in both subtle and profound ways. By practicing mindful and intentional observation, we can positively influence social outcomes, build stronger connections, and create environments where people thrive.

In the next chapter, we'll explore how ethical considerations and respect for privacy shape the balance between observation and freedom in the digital age.

CHAPTER 12:
THE ETHICS OF
OBSERVATION

Observation is a powerful tool, but with great power comes great responsibility. The ability to observe influences behavior, decisions, and outcomes, making it essential to consider the ethical implications of when, how, and why we observe. From parenting to workplace monitoring to technology and surveillance, the ethics of observation touch every aspect of our lives.

In this chapter, we'll explore the ethical dilemmas of observation, examine real-world examples, and provide guidance on how to balance observation with respect and empathy. By reflecting on these issues, we can ensure that observation is used as a force for good rather than a tool for harm.

• •

Why Ethics Matter in Observation

Observation is not inherently good or bad—it's the intent, context, and impact that determine its morality. Ethical observation respects:

1. **Privacy**:

    – 	Everyone has a right to personal space, free from unwanted intrusion.

2. **Consent**:

    – 	Observation should be transparent and agreed upon, not

hidden or manipulative.

3. **Dignity**:

   –   Observation should honor the humanity and individuality of the observed.

When these principles are ignored, observation can lead to exploitation, distrust, and harm.

• •

Ethical Dilemmas in Observation

1. Surveillance

Surveillance, whether by governments, corporations, or individuals, raises significant ethical concerns:

• **Benefits**:

   –   Surveillance can enhance safety and security, such as monitoring for criminal activity or ensuring workplace compliance.

• **Drawbacks**:

   –   Excessive or secretive surveillance infringes on privacy and can create a culture of fear and mistrust.

Example:

•   Many cities use facial recognition technology to monitor public spaces. While this can deter crime, it also raises concerns about misuse, data breaches, and the potential for targeting specific groups unfairly.

• •

2. Parenting

Parents naturally observe their children to guide and protect them, but ethical challenges arise when observation becomes overbearing:

• **Positive Observation**:

   –   Parents who watch attentively and empathetically foster trust and growth in their children.

- **Negative Observation**:

   – Constant monitoring, such as tracking every movement with GPS or reading private messages, can undermine a child's sense of independence and privacy.

Example:
- A parent uses an app to monitor their teenager's location. While the intention is safety, overuse of this technology can erode trust, leading the teen to feel micromanaged and resentful.

3. Workplace Monitoring

Employers often observe their employees to ensure productivity and compliance, but this can cross ethical boundaries:

- **Fair Observation**:

   – Tracking work output transparently and with employee consent can boost accountability.

- **Excessive Monitoring**:

   – Using hidden cameras or software to monitor every keystroke creates an atmosphere of distrust and invades employees' personal space.

Example:
- A company installs software to monitor employees working remotely. While it ensures productivity, employees may feel demoralized and disengaged if they believe their every move is scrutinized.

4. Technology and Data Collection

In the digital age, observation often occurs through data collection:

- **The Convenience Tradeoff**:

   – Services like personalized ads or recommendations

rely on observing user behavior, often without explicit consent.

- **Privacy Concerns**:

    – Data breaches and misuse of collected information can lead to identity theft, manipulation, or discrimination.

Example:

- Social media platforms track user activity to serve targeted ads. While this benefits advertisers and enhances user experience, many users are unaware of how their data is used, raising ethical concerns about transparency and consent.

• •

Balancing Observation with Respect and Empathy

Ethical observation requires finding a balance between the benefits of observation and the rights of the observed. Here are practical ways to approach observation with respect and empathy:

1. Be Transparent

- Clearly communicate why observation is taking place, who is observing, and how the information will be used.

    – Example: A workplace policy that outlines monitoring practices and seeks employee input fosters trust and collaboration.

2. Seek Consent

- Whenever possible, obtain consent before observing. This ensures that the observed feel respected and involved.

    – Example: Parents can discuss tracking apps with their teenagers, agreeing on when and why they will be used.

3. Limit Intrusion

- Observe only as much as necessary to achieve the intended goal. Avoid unnecessary or invasive methods.

    – Example: Instead of constantly monitoring employee activity, use periodic check-ins to assess progress.

4. Show Empathy

- Consider the perspective of the observed. How would you feel in their position? Empathy builds trust and reduces the risk of harm.

  - Example: A teacher observing a struggling student can offer support and encouragement rather than judgment.

5. Use Observation to Empower

- Observation should uplift and protect rather than control or manipulate. Focus on how the information gained can benefit both parties.

  - Example: A parent observing their child's interests can encourage them to pursue hobbies and passions.

• •

Relatable Examples of Ethical Observation

1. **A Fair Approach to Surveillance**:

   - A city uses cameras in public spaces but ensures the data is stored securely, only accessed for crime prevention, and shared transparently with the community.

2. **Parenting with Respect**:

   - A parent observes their child's online activity but discusses boundaries openly, explaining the risks of the internet while respecting their child's growing independence.

3. **Technology with Consent**:

   - A health app collects user data to provide personalized insights, but it clearly explains its privacy policy and allows users to opt out of data sharing.

• •

The Consequences of Unethical Observation

When observation lacks respect or empathy, it can lead to:

1. **Erosion of Trust**:

   –   Relationships, whether personal or professional, suffer when people feel spied on or manipulated.

2. **Loss of Autonomy**:

   –   Excessive observation stifles creativity and independence, as people feel they must conform to expectations.

3. **Social Harm**:

   –   Unchecked surveillance or data collection can disproportionately affect marginalized groups, perpetuating inequality and discrimination.

• •

Summary

Observation is a powerful tool that must be wielded responsibly. Whether in parenting, workplaces, or technology, ethical observation requires transparency, consent, and empathy. By respecting privacy and balancing the benefits of observation with its potential harms, we can create environments where observation empowers rather than exploits.

As we continue exploring the observer effect, the next chapter will delve into how observation can create intentional positive changes in your own life and relationships.

CHAPTER 13: BECOMING A CONSCIOUS OBSERVER

Observation shapes reality, whether in the quantum world or everyday life. But to harness its power fully, you must learn to observe consciously—with intention, mindfulness, and clarity. Conscious observation is about more than noticing what's happening; it's about actively engaging with your thoughts, feelings, and surroundings in ways that create positive change.

In this chapter, we'll explore techniques for becoming a more mindful observer, provide exercises to deepen your awareness, and share inspiring examples of people who transformed their lives through conscious observation.

* *

What Does It Mean to Be a Conscious Observer?

A conscious observer is someone who:

1. **Pays Attention**:

   – Fully notices the details of their internal and external world.

2. **Remains Nonjudgmental**:

   – Observes without rushing to label or react to what they see.

3. **Engages Intentionally**:

–   Chooses how to respond to observations in a way that aligns with their values and goals.

Conscious observation is the antidote to autopilot living. It empowers you to step out of reactive patterns and make deliberate choices.

• •

Techniques to Cultivate Conscious Observation

1. Practice Mindfulness

Mindfulness is the practice of bringing your full attention to the present moment. It's a cornerstone of conscious observation.

• **How to Practice**:

–   Take a few minutes each day to focus on your breath. Notice the sensation of air entering and leaving your body without trying to change it.

–   When thoughts arise, observe them without judgment and gently return your focus to your breath.

2. Journaling

Writing about your thoughts and experiences helps clarify your observations and deepen self-awareness.

• **How to Practice**:

–   Set aside 10 minutes each evening to journal about your day. Reflect on:

•   What did I observe today in myself and others?

•   How did I respond to those observations?

•   What patterns am I noticing?

3. Body Scans

Your body holds valuable information about your emotional and physical state. A body scan helps you tune into these signals.

• **How to Practice**:

– Sit or lie down in a quiet space. Slowly bring your attention to each part of your body, starting at your toes and moving upward. Notice any tension, pain, or relaxation without trying to change it.

4. Active Listening

Conscious observation extends to your interactions with others. Active listening helps you fully engage in conversations.

- **How to Practice**:

– When someone is speaking, focus entirely on their words, tone, and body language. Resist the urge to interrupt or plan your response while they're talking.

• •

Exercises to Deepen Awareness

Exercise 1: Observing Thoughts

This exercise helps you become aware of your mental patterns.

1. **Set a Timer**:

– Spend five minutes sitting quietly with your eyes closed.

2. **Watch Your Thoughts**:

– Imagine your thoughts as clouds drifting across the sky. Don't try to stop them; just notice their shape, speed, and content.

3. **Reflect**:

– After the timer ends, jot down what you noticed. Were there recurring themes? Did certain thoughts trigger emotions?

• •

Exercise 2: Sensory Awareness Walk

This exercise sharpens your observation of the external world.

1. **Take a Walk**:

– Go for a 10–15-minute walk in a familiar place.

2. **Engage Your Senses**:

– Focus on one sense at a time:

  • Sight: Notice colors, patterns, and movement.

  • Sound: Listen for subtle and distant sounds.

  • Smell: Identify scents in the air.

  • Touch: Feel textures underfoot or in your hands.

3. **Reflect**:

– How did focusing on your senses change your experience of the walk?

. . . .

Exercise 3: Observing Emotions

This exercise helps you understand your emotional responses.

1. **Pause During Emotional Moments**:

– The next time you feel a strong emotion, pause for a moment before reacting.

2. **Label the Emotion**:

– Identify what you're feeling (e.g., anger, joy, sadness).

3. **Explore the Cause**:

– Ask yourself: What triggered this emotion? What thoughts or beliefs are fueling it?

4. **Choose Your Response**:

– Decide how to respond in a way that aligns with your values.

. . . .

Real-Life Examples of Transformation Through Awareness

1. Overcoming Anxiety

A young professional named Sarah struggled with anxiety during work presentations. Through mindfulness practice, she learned to observe her thoughts without judgment. By noticing patterns —like catastrophizing or self-doubt—she was able to reframe her thoughts and approach presentations with greater confidence. Over time, her anxiety diminished, and her performance improved.

. . . .

2. Strengthening Relationships

Mark, a father of two, realized he was often distracted during family time. By practicing active listening and mindfulness, he became more present with his children. Observing their needs and emotions deepened their connection and transformed his role as a parent.

. . . .

3. Breaking Negative Habits

Samantha wanted to quit smoking but struggled with cravings. By journaling, she observed that her cravings were strongest after stressful meetings. This awareness allowed her to replace the habit with deep breathing exercises, eventually helping her quit.

. . . .

Why Conscious Observation Works

1. **Breaks Automatic Patterns**:

    – Awareness creates a pause between trigger and reaction, allowing for intentional choice.

2. **Reinforces Positive Change**:

    – Observing progress, no matter how small, builds motivation and resilience.

3. **Enhances Empathy**:

    – Observing others with curiosity and openness fosters stronger connections and reduces conflict.

. . . .

Practical Applications in Daily Life

1. **Morning Routine**:

   –   Start your day with a mindfulness exercise to set a tone of awareness and intention.

2. **Workplace Observation**:

   –   Practice observing your coworkers' nonverbal cues to improve communication and collaboration.

3. **Evening Reflection**:

   –   End your day by journaling or meditating on what you observed and learned.

. . . .

Summary

Becoming a conscious observer is a transformative practice that empowers you to live with greater awareness, intention, and connection. Through techniques like mindfulness, journaling, and active listening, you can deepen your understanding of yourself and others. By observing your thoughts, feelings, and surroundings, you unlock the power to break negative patterns, strengthen relationships, and create a life aligned with your values.

In the next chapter, we'll explore how to apply the principles of conscious observation to relationships, uncovering the profound impact it has on social dynamics and emotional bonds.

CHAPTER 14: QUANTUM LIVING – SHAPING YOUR REALITY

Quantum mechanics teaches us that observation shapes reality, and this principle is not limited to subatomic particles. By applying quantum principles in daily life, you can become an intentional creator of your experiences. Through conscious attention and focus, you can influence your work, relationships, and creativity to align with your goals and values.

This chapter explores how to integrate quantum-inspired concepts into practical living. We'll provide actionable tips and guide you through an activity to set daily intentions, helping you observe their tangible impact on your actions and outcomes.

• •

What Is Quantum Living?

Quantum living is the practice of applying principles from quantum mechanics—such as observation, focus, and interconnectedness—to your daily life. At its core, it's about recognizing that:

1. **Your Attention Is Powerful**:

    –   Like the observer effect in quantum physics, where observation influences particles, your attention shapes the outcomes you experience.

2. **Possibilities Are Limitless**:

- Quantum mechanics reveals that reality exists as probabilities until observed. Similarly, your life is filled with potential paths waiting for your focus to bring them to fruition.

3. **Interconnectedness Is Key**:

- Just as quantum particles are connected through entanglement, your thoughts, actions, and relationships are deeply intertwined, influencing one another.

• •

Practical Tips for Shaping Your Reality

1. Harness Attention to Influence Work

In your professional life, focused attention can help you achieve clarity, productivity, and success.

• **Eliminate Distractions**:

- Just as a particle's behavior is determined by observation, your productivity is determined by where you focus your energy. Turn off notifications, create a clutter-free workspace, and dedicate uninterrupted time to priority tasks.

• **Visualize Success**:

- Before starting a project or presentation, take a few moments to visualize a positive outcome. Imagine the steps flowing smoothly and the satisfaction of completing your work.

• **Celebrate Small Wins**:

- Observing and acknowledging progress reinforces motivation and momentum.

• •

2. Strengthen Relationships Through Conscious Observation

Relationships flourish when you focus your attention on

understanding and connecting with others.

- **Be Present**:

  - In conversations, practice active listening. Observe the other person's tone, body language, and emotions to deepen understanding.

- **Express Appreciation**:

  - Take time to observe and acknowledge the positive qualities of the people in your life. A simple compliment or expression of gratitude can strengthen bonds.

- **Redirect Negative Patterns**:

  - If a relationship dynamic feels strained, observe the habitual interactions contributing to tension. Use this awareness to introduce constructive communication.

. . . .

3. Spark Creativity by Tapping Into Possibilities

Creativity thrives in a space of openness and exploration, much like the quantum state of probabilities.

- **Set an Intention for Creative Flow**:

  - Start each creative session with a clear intention, such as "Today, I will explore bold ideas" or "I will embrace curiosity."

- **Experiment Without Judgment**:

  - Like a quantum particle exploring multiple paths, allow yourself to brainstorm freely without worrying about perfection.

- **Observe and Reflect**:

  - Review your creative process regularly to identify patterns, strengths, and areas for growth.

. . . .

Activity: Set a Daily Intention

This activity will help you practice quantum living by using focused intention to shape your day.

• •

STEP 1: CHOOSE A DAILY INTENTION

·   Think about an area of your life you'd like to improve or influence. Set a clear, positive intention for the day, such as:

- "I will approach my tasks with focus and enthusiasm."

- "I will be fully present in my interactions."

- "I will embrace creativity and curiosity."

· ·

STEP 2: WRITE IT DOWN

- Writing solidifies your intention and makes it feel tangible. Place it somewhere visible, like on a sticky note or in your planner.

· ·

STEP 3: OBSERVE YOUR ACTIONS

- Throughout the day, consciously observe how your intention shapes your behavior and choices. Ask yourself:

  - Am I acting in alignment with my intention?

  - How is my focus influencing my actions?

• •

STEP 4: REFLECT IN THE EVENING

- At the end of the day, spend a few minutes reflecting on the impact of your intention:

  – Did you notice any changes in your mindset or actions?

  – Were there moments when you felt particularly aligned with your intention?

  – What can you adjust or improve for tomorrow?

• •

Real-Life Examples of Quantum Living

1. Career Transformation

Anna, a graphic designer, struggled with creative blocks and self-doubt. By setting a daily intention to approach her work with curiosity and openness, she began to notice subtle shifts. Her focus on exploration led to innovative designs and renewed confidence in her abilities.

• •

2. Strengthened Relationships

David felt disconnected from his teenage son. He started setting an intention each morning to listen actively and engage without judgment. Over time, this conscious attention helped rebuild trust and deepen their connection.

• •

3. Boosted Productivity

Sophia, a small business owner, implemented focused attention techniques, such as visualizing her day's priorities and celebrating small wins. Her productivity soared, and she found greater satisfaction in her work.

• •

Why This Works: The Science Behind Quantum Living

1. **The Observer Effect**:

   – Just as observing particles influences their state, focusing your attention directs your energy and actions toward desired outcomes.

2. **Neuroplasticity**:

   – Intention and focus reshape the brain's neural pathways, reinforcing positive habits and mindsets.

3. **Energy Amplification**:

   – Attention acts like a magnifying glass, intensifying the impact of your thoughts and actions on your reality.

• •

Practical Applications in Daily Life

1. **Morning Routine**:

   – Start each day with a moment of reflection to set your intention and focus your energy.

2. **Mindful Breaks**:

   – Pause periodically during the day to observe whether you're aligned with your goals and adjust as needed.

3. **Gratitude Practice**:

   – End your day by observing and appreciating the ways your intentions influenced your reality.

• •

Summary

Quantum living bridges the principles of quantum mechanics

and daily life, empowering you to shape your reality through intentional focus and observation. By setting daily intentions and observing their impact, you can align your actions with your values, foster meaningful relationships, and unlock your creative potential.

In the next chapter, we'll explore how quantum principles can be applied to rewriting your personal narrative, allowing you to transform limiting beliefs and create a new vision for your life.

CHAPTER 15: REWRITING YOUR STORY THROUGH OBSERVATION

Your life is shaped by the stories you tell yourself. These internal narratives—about who you are, what you're capable of, and how the world works—act as the framework for your actions, decisions, and experiences. But what happens when those stories are limiting or disempowering? Through the power of observation, you can identify and rewrite these narratives, transforming your perspective and unlocking new possibilities.

In this chapter, we'll explore how to reframe negative stories into empowering ones, provide practical steps for observing and rewriting personal narratives, and share inspiring case studies of individuals who transformed their lives by shifting their perspectives.

• •

The Power of Stories

Stories are the lens through which we interpret the world and our place in it. While some stories empower us, others create barriers:

1. **Empowering Narratives**:

   – These stories highlight your strengths, resilience, and potential, motivating you to grow and succeed.

   – Example: "I've faced challenges before and come out

stronger, so I can handle this too."

2. **Limiting Narratives**:

- These stories focus on perceived failures, weaknesses, or external obstacles, keeping you stuck in self-doubt.

- Example: "I've always been bad at this, so there's no point in trying."

The good news is that your stories are not set in stone. By observing them consciously, you can reframe limiting narratives into ones that inspire and empower you.

• •

How Observation Helps Rewrite Stories

Observation creates space between your thoughts and your identity, allowing you to:

1. **Recognize Patterns**:

- By observing your inner dialogue, you can identify recurring themes and beliefs that influence your behavior.

2. **Question Assumptions**:

- Observation helps you challenge the validity of negative stories. Are they based on facts, or are they distorted by fear or past experiences?

3. **Introduce New Perspectives**:

- Through conscious observation, you can reframe old stories in ways that align with your goals and values.

• •

Practical Steps to Observe and Rewrite Personal Stories

STEP 1: IDENTIFY YOUR NARRATIVE

- Reflect on an area of your life where you feel stuck or unsatisfied. Ask yourself:

  – What story am I telling myself about this situation?

  – Examples:

    - "I'm not good at relationships."

    - "I'll never succeed in my career."

    - "I can't change because that's just who I am."

• •

STEP 2: OBSERVE WITHOUT JUDGMENT

- Spend a week observing your inner dialogue around this story. Use journaling to capture your thoughts:

    – What triggers this narrative?

    – How does it make you feel?

    – What actions or inactions does it lead to?

. . . .

STEP 3: CHALLENGE THE STORY

- Question the validity of your narrative:
    - Is this story absolutely true?
    - What evidence supports or contradicts it?
    - Could there be another way to interpret this situation?

• •

STEP 4: REFRAME THE NARRATIVE

- Rewrite your story in a way that acknowledges your strengths and opens possibilities:

    - Old Story: "I'm not good at relationships because I always mess things up."

    - Reframed Story: "I'm learning from my past experiences and becoming better at building meaningful connections."

. . . .

STEP 5: REINFORCE THE NEW STORY

- Practice repeating your new narrative daily. Write it down, say it aloud, or reflect on it during moments of doubt.

- Take small actions that align with your new story, reinforcing its validity over time.

• •

Case Studies: Transforming Lives Through Reframing

Case Study 1: From "I'm a Failure" to "I'm Resilient"

Michael struggled with feelings of failure after losing his job. His inner narrative revolved around the belief that he wasn't good enough. Through observation, he realized this story stemmed from childhood experiences of criticism. By reframing his story to focus on resilience—"I've overcome challenges before, and this is an opportunity to grow"—he regained confidence, started a new career, and rebuilt his self-worth.

• •

Case Study 2: From "I'm Stuck" to "I'm Evolving"

Sophia felt trapped in an unfulfilling relationship but believed she couldn't leave because she was "too dependent." Observing her thoughts revealed a pattern of self-doubt tied to past insecurities. By reframing her story as one of evolution—"I'm learning to stand on my own and prioritize my happiness"—she found the courage to make empowering decisions.

• •

Case Study 3: From "I'm Not Creative" to "I'm a Problem-Solver"

Lila, an engineer, told herself she lacked creativity, which made her hesitant to pursue a personal passion for writing. Through observation, she realized her definition of creativity was too narrow. By reframing her story—"My problem-solving skills are a form of creativity"—she began writing short stories and discovered a new outlet for self-expression.

• •

Tips for Rewriting Your Story

1. **Use Empowering Language**:

   –   Frame your narrative with words that emphasize growth, possibility, and resilience.

   –   Example: Replace "I can't" with "I'm learning how to."

2. **Focus on Progress, Not Perfection**:

   –   Highlight small steps and improvements rather than fixating on past mistakes or unachieved goals.

3. **Seek Support**:

   –   Share your new narrative with trusted friends or mentors who can reinforce and encourage your growth.

4. **Practice Gratitude**:

   –   Reflect on what you've learned and achieved, even in challenging situations. Gratitude helps shift your focus toward positivity.

• •

Activity: Observe and Rewrite a Limiting Story

STEP 1: CHOOSE A LIMITING NARRATIVE

- Identify one story you've been telling yourself that feels disempowering. Write it down.

STEP 2: OBSERVE THE NARRATIVE

- Spend a week observing when and how this story arises. Journal your observations:

    – What situations trigger the story?

    – How does it affect your emotions and actions?

STEP 3: REFRAME
THE STORY

- Rewrite the narrative in a way that acknowledges your strengths and opens new possibilities. Write your new story alongside the old one.

STEP 4: ACT ON THE NEW STORY

- Take one small action that aligns with your reframed narrative. For example:

  - If your new story is about building confidence, volunteer to lead a small project at work.

STEP 5: REFLECT AND ADJUST

- At the end of the week, review your progress. How did reframing the story change your perspective or behavior? Adjust your narrative as needed.

• •

Why Rewriting Your Story Matters

1. **Empowers Personal Growth**:

   – Shifting your perspective frees you from limiting beliefs and opens paths to growth.

2. **Aligns Actions with Values**:

   – A positive narrative encourages behaviors that align with your goals and aspirations.

3. **Builds Resilience**:

   – By focusing on strengths and possibilities, you develop the confidence to navigate challenges.

• •

Summary

The stories you tell yourself shape your reality, but they're not immutable. By observing your narratives, questioning their validity, and reframing them into empowering stories, you can transform your perspective and your life. Whether it's overcoming self-doubt, strengthening relationships, or pursuing new goals, the process of rewriting your story begins with conscious observation.

In the next chapter, we'll explore how to integrate the lessons of observation and intentional living into a lifelong practice, ensuring your journey of growth and transformation continues.

CHAPTER 16:
THE FUTURE OF
OBSERVATION

The observer effect has transformed our understanding of the quantum world and hinted at profound connections between observation, consciousness, and reality. As technology advances, particularly in quantum research, artificial intelligence (AI), and neuroscience, our ability to explore the nature of observation will expand in ways that could reshape our understanding of existence itself.

This chapter explores how emerging technologies like quantum computing, AI, and neuroscience intersect with the observer effect, speculating on their potential to unlock deeper truths about the universe and their implications for humanity's future.

• •

Technological Advancements in Quantum Research

1. Quantum Computing and Observation

Quantum computing represents a revolutionary step in harnessing the principles of quantum mechanics to solve problems far beyond the capabilities of classical computers.

- **Superposition and Entanglement**:

    – Quantum computers use qubits, which can exist in multiple states simultaneously (superposition), and entanglement, where the state of one qubit affects another instantaneously, regardless of distance.

- **The Role of Observation**:

- Observation in quantum computing isn't just a theoretical concept—it's central to how computations are processed. The act of measuring a qubit collapses its state, enabling the retrieval of results.

- **Future Implications**:

  - Quantum computing could help simulate and understand complex quantum systems, shedding light on the observer effect itself.

• •

2. Advanced Experimental Tools

New tools are enabling scientists to probe the quantum world with unprecedented precision:

- **Quantum Microscopes**:

  - These devices allow researchers to directly observe quantum phenomena at incredibly small scales, pushing the boundaries of what we know about particles and waves.

- **Time-Crystal Research**:

  - Time crystals, a recently discovered phase of matter, exhibit perpetual motion without energy input. They challenge classical physics and may reveal new aspects of the observer effect.

• •

Artificial Intelligence and the Observer Effect

1. AI as a New Kind of Observer

Artificial intelligence systems are already playing a role in observing and interpreting complex phenomena:

- **AI in Quantum Research**:

  - Machine learning algorithms analyze quantum experiments, identifying patterns and outcomes that might elude human observers.

- **Autonomous Observation**:

  – AI systems could become independent observers, measuring quantum states and influencing outcomes in ways we're only beginning to understand.

2. AI's Role in Consciousness Studies

AI is also being used to model and study human consciousness, potentially bridging the gap between quantum observation and neuroscience:

- **Simulating Observation**:

  – AI might simulate aspects of human perception and test hypotheses about the connection between observation and reality.

- **AI and Free Will**:

  – The study of decision-making in AI systems raises questions about whether machines can "observe" in a way that influences outcomes, mirroring the observer effect.

• •

Neuroscience and the Observer Effect

1. Understanding Consciousness

Neuroscience is advancing our understanding of how the brain generates awareness and perception, directly connecting to questions about the role of consciousness in the observer effect.

- **The Brain as an Observer**:

  – Studies using brain imaging technologies like fMRI and EEG reveal how the brain processes and interprets sensory information, offering clues about how observation shapes subjective reality.

- **Quantum Processes in the Brain**:

  – Some researchers, such as those exploring the Penrose-Hameroff Orch-OR theory, propose that quantum processes in brain microtubules could play a role in consciousness, linking neuroscience to quantum mechanics.

2. Enhanced Awareness Through Technology

Technological advances are allowing for enhanced observation of the self:

- **Brain-Computer Interfaces (BCIs)**:

  - BCIs, which allow direct communication between the brain and computers, could offer new ways to observe and influence our own thoughts and actions.

- **Neurofeedback**:

  - Real-time brain monitoring can help individuals become more aware of their mental states, enabling greater control over focus, mood, and decision-making.

• •

Speculation: How Technology Might Reshape
Our Understanding of Reality

1. Quantum-AI Synergy

The combination of AI and quantum computing could revolutionize how we study and interact with the quantum world:

- **Deeper Simulations**:

  - AI could model entire quantum systems, revealing how observation impacts complex phenomena like entanglement and superposition.

- **Discovering New Dimensions**:

  - Advanced technology might uncover hidden dimensions or layers of reality, offering new insights into the fabric of the universe.

2. Consciousness as a Field

Technological advancements may validate theories suggesting that consciousness is not localized to the brain but is a fundamental field of the universe:

- **Global Consciousness Studies**:

  – Experiments measuring collective human attention, such as those conducted by the Global Consciousness Project, could gain precision with AI and quantum sensors, revealing large-scale effects of observation on reality.

3. Reimagining Human Potential

If observation truly shapes reality, tools like BCIs and neurofeedback could amplify our ability to intentionally influence outcomes:

- **Directed Manifestation**:

  – Individuals might use enhanced observation techniques to consciously manifest desired states, bridging the gap between science and ancient practices like visualization and meditation.

• •

Implications for Humanity's Future

1. Ethical Challenges

The power of observation, combined with advanced technologies, raises ethical questions:

- **Privacy in a Hyper-Observed World**:

  – As observation becomes more precise and pervasive, how do we protect individual privacy?

- **Autonomy and Manipulation**:

  – Could the ability to shape reality through observation be exploited by powerful entities for control or manipulation?

2. Philosophical Shifts

Advances in quantum research, AI, and neuroscience may challenge our foundational beliefs:

- **Redefining Free Will**:

- If observation determines outcomes, what does this mean for our understanding of free will and personal responsibility?

- **The Nature of Reality**:

    - Discoveries in quantum mechanics and consciousness could lead to a new paradigm where reality is understood as a dynamic interplay between observers and the observed.

3. A Connected Humanity

Technological and scientific advancements could deepen our understanding of interconnectedness:

- **Collective Observation**:

    - If collective attention influences reality, humanity might unite to focus on shared goals, such as solving global challenges or fostering peace.

- **Spiritual and Scientific Integration**:

    - As science confirms ancient insights about observation and interconnectedness, it could bridge the gap between spirituality and technology, fostering a more harmonious worldview.

• •

Summary

The future of observation lies at the intersection of quantum mechanics, AI, and neuroscience. These advancements hold the potential to reshape our understanding of reality, consciousness, and human potential. While they bring incredible opportunities, they also demand careful ethical consideration and philosophical reflection.

As humanity continues to explore the mysteries of observation, we may find ourselves on the cusp of a profound shift—a new understanding of our role as active participants in the universe, capable of shaping reality through the power of focus, attention, and intention.

In the next chapter, we'll reflect on how the lessons of this book can be integrated into a lifelong practice of conscious living, empowering you to create a reality that aligns with your highest potential.

CHAPTER 17: OBSERVING THE GLOBAL STAGE

Observation shapes not only individual realities but also the collective experience of humanity. When groups of people focus their attention on a shared goal or issue, their collective observation influences the trajectory of culture, politics, and even the environment. In the quantum sense, collective observation can collapse possibilities into specific outcomes, but on a global scale, it manifests as the power of shared focus to create change.

In this chapter, we'll explore how collective observation impacts the world, using examples from global movements, media influence, and environmental awareness. Finally, we'll challenge you to embrace your role as a conscious observer in the broader context of humanity's journey.

• •

How Collective Observation Shapes the World

1. **Shared Focus and Cultural Shifts**

   – When groups of people direct their attention to a particular issue, it gains momentum, influencing societal norms and values.

   – Example:

     • The civil rights movement in the United States captured the world's attention, leading to profound cultural and political changes.

2. **The Role of Media in Shaping Attention**

–   Media, as a powerful tool of observation, directs collective focus, shaping public opinion and priorities.

–   Example:

   •   Coverage of climate change has increased global awareness, prompting action by governments, corporations, and individuals.

3. **The Butterfly Effect of Attention**

–   Small acts of observation can ripple outward, influencing large-scale events.

–   Example:

   •   A viral social media post about injustice can spark worldwide protests and policy changes.

• •

Examples of Collective Observation

1. Global Movements

Global movements often begin with a small group of people drawing attention to an issue, but they succeed when collective observation amplifies their message.

•   **Black Lives Matter**:

   –   What began as a grassroots effort to highlight systemic racism gained international attention through media and social platforms, driving conversations and policy reforms worldwide.

•   **Fridays for Future**:

   –   Greta Thunberg's solo protests for climate action grew into a global movement as millions of people focused their attention on the urgency of environmental issues.

• •

2. Media Influence and Shared Narratives

Media acts as a lens that shapes how the world is observed and understood.

- **The Power of Headlines**:

    – News outlets often frame stories in ways that influence public perception. For example, media coverage of refugee crises can evoke empathy or fear, depending on the narrative presented.

- **Social Media Echo Chambers**:

    – Platforms like Twitter and Facebook amplify certain perspectives through algorithms, creating collective focus around trending topics. While this can raise awareness, it can also deepen divisions by reinforcing biases.

• •

3. Environmental Observation

Human observation of the environment plays a dual role: raising awareness and influencing behavior.

- **Citizen Science**:

    – Individuals observing and reporting environmental data, such as bird migrations or pollution levels, contribute to global understanding and conservation efforts.

- **Focus on Sustainability**:

    – Increased awareness of plastic pollution, fueled by viral campaigns and documentaries, has led to changes in consumer habits and corporate policies.

• •

The Role of Conscious Observers in Shaping the Future

As a conscious observer, you have the power to influence the global stage in meaningful ways:

1. **Where You Focus Matters**:

    – The issues you choose to observe and discuss shape

public discourse and priorities. By focusing on solutions rather than problems, you can drive positive change.

2. **Participation Amplifies Impact**:

   – Joining movements, sharing information, or simply engaging in thoughtful conversations can magnify the effects of collective observation.

3. **Balancing Influence with Critical Thinking**:

   – While observing global issues, maintain awareness of biases in media and social narratives. Seek diverse perspectives to form a well-rounded understanding.

• •

Practical Steps for Becoming a Conscious Observer on the Global Stage

1. Engage with Media Mindfully

• **Diversify Your Sources**:

  – Follow news from multiple outlets with varying perspectives to avoid echo chambers.

• **Fact-Check Information**:

  – Verify the credibility of sources before sharing content or forming opinions.

2. Support Causes Aligned with Your Values

• **Raise Awareness**:

  – Use your voice and platforms to highlight issues that matter to you, whether it's climate action, human rights, or social justice.

• **Contribute to Solutions**:

  – Volunteer, donate, or advocate for initiatives that align with your values.

3. Practice Empathy in Observation

• **Listen to Understand**:

–   When engaging with global issues, consider the perspectives of those directly affected. Empathy deepens your connection to the cause.

- **Avoid Judgment**:

    –   Observe without rushing to conclusions or blaming others. Compassion fosters collaboration and unity.

• •

Activity: Focused Collective Attention

This activity helps you experience the power of shared focus in creating change.

STEP 1: CHOOSE AN ISSUE

- Select a global or local issue that resonates with you, such as climate change, education inequality, or mental health awareness.

STEP 2: RESEARCH AND REFLECT

- Learn about the issue from multiple perspectives. Reflect on:
  - What is the core problem?
  - What solutions are being proposed?

STEP 3: DIRECT YOUR ATTENTION

• Spend a week focusing your energy on this issue. Actions might include:

– Sharing information on social media.

– Writing to local leaders or policymakers.

– Participating in a related event or movement.

STEP 4: OBSERVE THE RIPPLE EFFECT

- Reflect on how your focus influenced your thoughts, conversations, and actions. Did others engage with the issue because of your efforts?

• •

Why Collective Observation Matters

The observer effect demonstrates that attention changes outcomes. On a global scale, collective observation has the potential to:

1. **Shift Priorities**:

   – When enough people focus on an issue, it becomes impossible to ignore, forcing leaders and institutions to act.

2. **Create Unity**:

   – Shared attention fosters a sense of connection and solidarity, reminding us of our shared humanity.

3. **Empower Change**:

   – By observing problems with an eye toward solutions, humanity can create a brighter future.

• •

Summary

Collective observation is a powerful force that shapes culture, politics, and the environment. By understanding the role of shared focus and participating as a conscious observer, you can contribute to meaningful change. Whether through raising awareness, engaging with global movements, or practicing empathy, your attention has the potential to influence humanity's trajectory.

In the next chapter, we'll integrate the lessons of observation into a unified vision, offering a framework for embracing the power of attention to shape a harmonious and interconnected future.

CONCLUSION: YOUR ROLE AS THE OBSERVER

As we come to the end of this journey, it's time to reflect on the transformative power of observation and your unique role in shaping reality. Throughout this book, we've explored how attention influences the quantum world, your personal experiences, and the global stage. We've examined the profound implications of the observer effect, not only as a scientific phenomenon but as a guiding principle for life.

The act of observing is more than a passive process—it is a creative force. By choosing where and how you direct your attention, you hold the power to influence outcomes, inspire change, and manifest your vision of the future. Let's revisit the key lessons we've uncovered and envision how you can integrate them into your life.

• •

Recap of Key Lessons

1. Observation Shapes Reality

- **In the Quantum World**:

    - At the subatomic level, particles exist as probabilities until observed, collapsing into specific states. This principle demonstrates the profound relationship between attention and existence.

- **In Daily Life**:

- Your attention directs energy. What you focus on—your habits, relationships, and goals—grows and transforms.

2. The Observer Effect in Personal Transformation

- **Self-Reflection**:

  - Observing your thoughts and emotions brings clarity and helps break automatic patterns.

- **Reframing Narratives**:

  - By observing and rewriting limiting beliefs, you can create empowering stories that align with your aspirations.

3. Collective Observation and Global Impact

- **Shared Focus**:

  - When people unite their attention on common goals, they shape culture, politics, and the environment.

- **Responsibility**:

  - As a conscious observer, your choices and focus influence not only your life but the collective reality of humanity.

4. The Ethics of Observation

- **Mindfulness**:

  - Observing with respect and empathy ensures that your focus uplifts and empowers rather than invades or manipulates.

- **Intentionality**:

  - Aligning your attention with your values creates harmony between observation and action.

• •

Embracing the Power of Observation

The ability to observe is one of humanity's greatest gifts. It allows us to:

1. **Discover**:

- Observation reveals the hidden patterns and possibilities in the world around us.

2. **Create**:

- By focusing our attention, we can turn potential into reality, whether through personal growth, creative endeavors, or social change.

3. **Connect**:

- Observing others with empathy and presence deepens relationships and fosters understanding.

• •

Your Role as a Conscious Observer

As an observer, you are both a participant in and a co-creator of reality. Every choice you make about where to focus your attention carries the potential to shape the world. Here are some ways to embrace your role:

1. **Be Present**:

- Mindful observation begins with being fully engaged in the moment. Notice the details of your thoughts, feelings, and environment.

2. **Focus on Growth**:

- Direct your attention toward opportunities, solutions, and strengths. Let go of distractions that pull you away from your purpose.

3. **Inspire Others**:

- Your observation creates ripples. By living intentionally, you encourage others to do the same, amplifying your impact on the world.

• •

A Lasting Thought: Your Attention Is the Most Powerful Force in the Universe

The universe is a canvas of infinite possibilities, and your attention is the brush. Each moment offers you the opportunity to observe, choose, and create. Whether you are shaping your personal story, building connections with others, or contributing to the collective future, your focus holds extraordinary power. Remember:

- What you choose to see defines what you experience.

- Where you direct your energy determines what grows.

- Who you become as an observer shapes the reality you live in.

The observer effect is more than a scientific principle—it is a call to action. Embrace your role as the conscious creator of your life and the world around you. The future is not a fixed destination; it is a dynamic interplay between your attention and the universe's infinite potential.

• •

FINAL CALL TO ACTION

As you close this book, ask yourself:

- What do I want to focus on today?

- How can I use my attention to create the reality I desire?

- What role will I play as an observer in shaping the future?

Let your answers guide you toward a life of intention, connection, and purpose. The universe is waiting to respond to your observation. Your journey begins now.

www.ingramcontent.com/pod-product-compliance
Lightning Source LLC
Chambersburg PA
CBHW071508220526
45472CB00003B/951